◆ ◆ ◆

AF442058

Copyright ©2020 ARNOLD KUNTZ PH.D

CONTENTS

ARNOLD KUNTZ PH.D

CONCLUSION

INTRODUCTION

Hydraulics is mechanical function that operates through the force of liquid pressure. In hydraulics-based systems, mechanical movement is produced by contained, pumped liquid, typically through cylinders moving pistons. Hydraulics is a component mechatronics, which combines mechanical, electronics and software engineering in the designing and manufacturing of products and processes. Simple hydraulic systems include aqueducts and irrigation systems that deliver water, using gravity to create water pressure. These systems essentially use water's own properties to make it deliver itself. More complex hydraulics use a pump to pressurize liquids (typically oils), moving a piston through a cylinder as well as valves to control the flow of oil. A log splitter is a single-piston hydraulic machine that uses a valve at either end of the cylinder that allows the pistons to be moved by the pressurized liquid, driving a wedge to force wood into smaller pieces and return to a home position. Force multiplication can be created by using a cylinder with a smaller diameter to push a larger piston in a larger cylinder. Often, there will be a number of pistons. Industrial equipment such as backhoes often use a number of cylinders to move different parts. Electronic controls are generally used for these more complicated setups on large, powerful equipment.

Hydraulics are similar to pneumatic systems in function. Both systems use fluids but, unlike pneumatics, hydraulics use liquids rather than gasses. Hydraulics systems are capable of greater pressures: up to 10000 pounds per square inch (psi) vs about 100 psi in pneumatics systems. This pressure is due to the incompressibility of liquids which enables greater power transfer with increased efficiency as energy is not lost to compression, except in the case where air gets into hydraulic lines. Fluids used in hydraulics may lubricate, cool and transmit power as well. Pneumatics, being less multifaceted, require oil lubrication separately, which can be messy with air pressure. Pneumatics are simpler in design and to control, safer (with less risk of fire) and more reliable, partially as the compressibility of the gas-absorbing shock can protect the mechanism.

Hydraulics (from Greek: Υδραυλική) is a technology and applied science using engineering, chemistry, and other sciences involving the mechanical properties and use of liquids. At a very basic level, hydraulics is the liquid counterpart of pneumatics, which concerns gases. Fluid mechanics provides the theoretical foundation for hydraulics, which focuses on the applied engineering using the properties of fluids. In its fluid power applications, hydraulics is used for the generation, control, and transmission of power by the use of pressurized liquids. Hydraulic topics range through some parts of science and most of engineering modules, and cover concepts such as pipe flow, dam design, fluidics and fluid control circuitry. The principles of hydraulics are in use naturally in the human body within the vascular system and erectile tissue. Free surface hydraulics is the branch of hydraulics dealing with free surface flow, such as occurring in rivers, canals, lakes, estuaries and seas. Its sub-field open-channel flow studies the flow in open channels.

AN OVERVIEW OF HYDRAULIC SYSTEMS

The purpose of a specific hydraulic system may vary, but all hydraulic systems work through the same basic concept. Defined simply, hydraulic systems function and perform tasks through using a fluid that is pressurized. Another way to put this is the pressurized fluid makes things work.

The power of liquid fuel in hydraulics is significant and as a result, hydraulic are commonly used in heavy equipment. In a hydraulic system, pressure, applied to a contained fluid at any point, is transmitted undiminished. That pressurized fluid acts upon every part of the section of a containing vessel and creates force or power. Due to the use of this force, and depending on how it's applied, operators can lift heavy loads, and precise repetitive tasks can be easily done. Marvelously versatile, hydraulic systems are dynamic, yet relatively straightforward in how they work. Let's look at some applications and a few basic components found in hydraulic systems.

HYDRAULIC CIRCUITS

Transporting liquid through a set of interconnected discrete components, a hydraulic circuit is a system that can control where fluid flows (such as thermodynamic systems), as well as control fluid pressure (such as hydraulic amplifiers). The system of a hydraulic circuit works similar to electric circuit theory, using linear and discrete elements. Hydraulic circuits are often applied in chemical processing (flow systems).

HYDRAULIC PUMPS

Mechanical power is converted into hydraulic energy using the flow and pressure of a hydraulic pump. Hydraulic pumps operate by creating a vacuum at a pump inlet, forcing liquid from a reservoir into an inlet line, and to the pump. Mechanical action sends the liquid to the pump outlet, and as it does, forces it into the hydraulic system. This is an example of Pascal's Law, which is foundational to the principle of hydraulics. According to Pascal's Law, "A pressure change occurring anywhere in a confined incompressible fluid is transmitted throughout the fluid such that the same change occurs everywhere."

HYDRAULIC MOTORS

The conversion of hydraulic pressure and flow into torque (or a twisting force) and then rotation is the function of a hydraulic motor, which is a mechanical actuator. The use of these is quite adaptable. Along with hydraulic cylinders and hydraulic pumps, hydraulic motors can be united in a hydraulic drive system. Combined with hydraulic pumps, the hydraulic motors can create hydraulic transmissions. While some hydraulic motors run on water, the majority in today's business operations are powered by hydraulic fluid, as the ones in your business likely are.

HYDRAULIC CYLINDERS

A hydraulic cylinder is a mechanism that converts energy stored in the hydraulic fluid into a force used to move the cylinder in a linear direction. It too has many applications and can be either single acting or double acting. As part of the complete hydraulic system, the cylinders initiate the pressure of the fluid, the flow of which is regulated by a hydraulic motor.

HYDRAULIC ENERGY AND SAFETY

Hydraulics present a set of hazards to be aware of, and for that reason safety training is required.

Remember, the purpose of hydraulic systems is to create motion or force. It's a power source, generating energy. Don't underestimate hydraulic energy in your safety program. It is small but mighty in force. And like any force, it can do great good or great harm. In the workplace, that translates to a potential hazard source, especially if uncontrolled. Hydraulic energy is subject to OSHA's Lockout/Tagout rules, along with electrical energy and other similar hazard sources. Be sure to train workers about the hazards of uncontrolled hydraulic energy, especially during maintenance, and the need for lockout/tagout. If neglected in procedures or forgotten when servicing equipment, uncontrolled hydraulic energy can have devastating results. Failure to control hydraulic energy frequently causes crushing events, amputations, and lacerations to exposed workers.

Therefore, like other energy sources, hydraulic energy must be controlled, using an appropriate energy isolating device that prevents a physical release of energy. There are also systems that require the release of stored hydraulic energy to relieve pressure. And also, those engaged in lockout/tagout, must also verify the release of stored hydraulic energy/pressure (usually indicated by zero pressure on gauges) prior to working on equipment. Also, workers need training which must explain the hazard potential and clearly detail methods to prevent injury. All employees who are authorized to lockout machines or equipment and perform the service and maintenance operations need to be trained in recognition of applicable hazardous energy sources in the workplace, the type and magnitude of energy found in the workplace, and the means and methods of isolating and/or controlling the energy."

You should be very familiar with any equipment in your business that creates hydraulic energy to ensure your workers are adequately protected through well-detailed procedures and training. And of course, your LO/TO program should echo your procedures, and list sources of workplace hydraulic energy devices. (Don't forget to perform at least annual reviews of the program and procedures to ensure you catch any changes or deficiencies.) Again, it's critical anyone involved with hydraulic systems is properly trained. Don't neglect that aspect. It's important to understand the principles of these systems, not only for servicing and maintenance, but also to understand the ways the hydraulic systems function to avoid injuries and accidents.

HYDRAULICS: FLUID MECHANICS

Hydraulics, branch of science concerned with the practical applications of fluids, primarily liquids, in motion. It is related to fluid mechanics (q.v.), which in large part provides its theoretical foundation. Hydraulics deals with such matters as the flow of liquids in pipes, rivers, and channels and their confinement by dams and tanks. Some of its principles apply also to gases, usually in cases in which variations in density are relatively small. Consequently, the scope of hydraulics extends to such mechanical devices as fans and gas turbines and to pneumatic control systems.

Liquids in motion or under pressure did useful work for man for many centuries before French scientist-philosopher Blaise Pascal and Swiss physicist Daniel Bernoulli formulated the laws on which modern hydraulic-power technology is based. Pascal's law, formulated in about 1650, states that pressure in a liquid is transmitted equally in all directions; i.e., when water is made to fill a closed container, the application of pressure at any point will be transmitted to all sides of the container. In the hydraulic press, Pascal's law is used to gain an increase in force; a small force applied to a small piston in a small cylinder is transmitted through a tube to a large cylinder, where it presses equally against all sides of the cylinder, including the large piston. Bernoulli's law, formulated about a century later, states that energy in a fluid is due to elevation, motion, and pressure, and if there are no losses due to friction and no work done, the sum of the energies remains constant. Thus, velocity energy, deriving from motion, can be partly converted to pressure energy by enlarging the cross section of a pipe, which slows down the flow but increases the area against which the fluid is pressing. Until the 19th century it was not possible to develop velocities and pressures much greater than those provided by nature, but the invention of pumps brought a vast potential for application of the discoveries of Pascal and Bernoulli. In 1882 the city of London built a hydraulic system that delivered pressurized water through street mains to drive machinery in factories. In 1906 an important advance in hydraulic techniques was made when an oil hydraulic system was installed to raise and control the guns of the USS "Virginia." In the 1920s, self-contained hydraulic units consisting of a pump, controls, and motor were developed, opening the way to applications in machine tools, automobiles, farm and earth-moving machinery, locomotives, ships, airplanes, and spacecraft.

In hydraulic-power systems there are five elements: the driver, the pump, the control valves, the motor, and the load. The driver may be an electric motor or an engine of any type. The pump acts mainly to increase pressure. The motor may be a counterpart of the pump, transforming hydraulic input into mechanical output. Motors may produce either rotary or reciprocating motion in the load. The growth of fluid-power technology since World War II has been phenomenal. In the operation and control of machine tools, farm machinery, construction machinery, and mining machinery, fluid power can compete successfully with mechanical and electrical systems (see fluidics). Its chief advantages are flexibility and the ability to multiply forces efficiently; it also provides fast and accurate response to controls. Fluid power can provide a force of a few ounces or one of thousands of tons.

Hydraulic-power systems have become one of the major energy-transmission technologies utilized by all phases of industrial, agricultural, and defense activity. Modern aircraft, for example, use hydraulic systems to activate their controls and to operate landing gears and brakes. Virtually all missiles, as well as their ground-support equipment, utilize fluid power. Automobiles use hydraulic-power systems in their transmissions, brakes, and steering mechanisms. Mass production and its offspring, automation, in many industries have their foundations in the utilization of fluid-power systems.

FLUIDICS

Fluidics, the technology of using the flow characteristics of liquid or gas to operate a control system. One of the newest of the control technologies, fluidics has in recent years come to compete with mechanical and electrical systems. Although fluidic principles are fairly old, it was not until about 1960 that researchers attempted to use fluidics commercially. The demand for reliable controls in space research stimulated progress. In the 1930s Henri Coandă, a Romanian scientist, described what is now known as the Coandă effect, a major contribution to fluidic technology. He observed that as a free jet emerges from a jet nozzle the stream will tend to follow a nearby curved or inclined surface. It also "attaches" itself to and flows along this surface if the curvature or angle of inclination is not too sharp. Coandă explained this tendency as being caused by the jet stream's entraining (picking up) nearby fluid molecules. When the supply of these molecules is limited by an adjacent surface, a partial vacuum develops between the jet and the surface. If the pressure on the other side of the jet remains constant, the partial vacuum, which is a lower pressure region, will force the jet to bend and attach itself to the wall.

Because fluidics is not as rapid as electronics, it is unlikely to compete in fields with ultrahigh-speed requirements. On the other hand, in many applications fluidics is advantageous. It is now possible to detect, interlock, and power complex operations by using air throughout a system. Controls can be installed by a competent fitter who might not be capable of dealing with electronic or electrical systems. The elimination of electrical contacts prevents a possible fire hazard. Pneumatic circuits require controls with simple interlocking, performed by air-piloted and mechanically operated control valves. Because it employs the same medium, fluidics is useful for sensitive detection and complex control as part of a pneumatic system. Combining hydraulics and fluidics is more complicated, however, because the same medium is not used. Yet since both systems require plumbing expertise, labor problems can be reduced. Power-output devices handling hydraulic pressures that respond to fluidic signals are available commercially. Fluidics has been applied to industrial problems on a wide scale, with no particular industry emerging as an obvious choice. Fluidics can operate in hazardous environments and sense objects by methods not available previously. Some typical applications include a weighing system that selects 10 different weights of raw material before machining; air-jet detection of delicate material (the roof lining of an automobile) that would be damaged by mechanical methods; and sonic detectors operating in the highly inflammable, contaminated area of a paint-spray booth. These detectors sense sound waves without disturbing the freshly painted surface.

PRINCIPLES OF OPERATION

Fluidic devices operate on either the digital principle (they are either "on" or "off") or the analog principle (the output of the device is continuously proportional to the input).

BASIC FLUIDIC CIRCUITS

In the case of an OR circuit, as shown in the figure, input a or b can produce an output signal, because each has a path through which the signal can flow to the output. This system is called logical (and digital) because no output is possible without an input. Either condition will satisfy the required output (the OR function, as an output, is produced whenever one or the other input is energized). When an AND circuit is involved, both inputs are required for an output because the flow from a or b alone, without a counterbalancing force, will go out one of the vents. If both are applied, they will collide, producing flow out of the center port marked a and b. This again is logical (and digital) because no output is possible unless both signals are applied. All conditions must be satisfied before an output is obtained (the AND function) as an output signal will be produced only if an input is applied to both inputs a and b simultaneously. With a proportional circuit, fluid flowing from the supply will go out of the vent unless input a is applied. This is an analog effect because the output can be altered proportionally from minimum to maximum by varying the power of input a. Fluidic devices can thus produce both logic (digital) and analog (proportional) effects or functions. The OR and the AND are the most common logic functions. In most fluidic devices, low-value input pressures or flows can control higher output pressures or flows. This is what is meant by the term fluid amplifier. A supply of fluid entering a device becomes a stream forced to follow a chosen path through carefully designed internal shapes before giving an output. Input jets of far lower power are positioned to give the greatest possible effect on the stream, thereby controlling the output. Fluid amplifiers respond to very small fluid signals provided by such devices as temperature or velocity detectors, generally by input sensors attached to existing mechanical movements. The number of devices controlled by one similar device is called the fan-out ratio. For example, if the output of one device is so strong that it can switch four others at the same time, the fan-out ratio is four.

TECHNOLOGICAL DEVELOPMENTS

Among the more recent advances in fluidics is modular construction of circuits i.e., construction of combinations of components that can be readily fitted together to form whole systems. A motor governor system, for example, converts pulsating frequencies of air motor exhausts into pressure levels, which are then compared to preset values. The difference in pressure is amplified to provide speed regulation of the motor. Converting the frequency of ON/OFF pulses into progressively increasing or decreasing values is called digital proportional. Fluidic devices are stacked in layers to provide a common supply and interconnections.

Another significant development is the edge tone amplifier, which works very much like a musical instrument; air blown at a sharp wedge oscillates at very high frequencies (about 5,000 hertz) to produce an output that is virtually continuous. Frequency of oscillation (sound) is controlled mechanically or by varying the force of the air directed at the wedge. Sound detection is possible with laminar streams that can be made sensitive to certain sound frequencies. A beam of sound can span distances for detection without even the slight force exerted by an air jet.

HOW DO HYDRAULICS WORK?

You probably know that solids are typically impossible to squish. If you pick up a solid object like a pen or piece of wood and try to squeeze it, nothing's going to happen to the materials. They won't compress or squish. Liquid works in the same way. It is incompressible, meaning it won't squeeze when you apply pressure to it. It takes up the same amount of space as it did when pressure wasn't applied to it. Picture water in a syringe. If you cap the end of it with your finger and try to press down, neither the water nor the plunger will go anywhere.

Where hydraulics are concerned, that incompressibility is a major player in making them work. In that same syringe, if you press down on the plunger normally, you'll release the water at high speed through the narrow end, even if you didn't apply that much pressure. When you push down the plunger, you apply pressure to the water, which will try to escape however it can in this case, at high pressure through a very narrow exit. This application shows us that we can multiply force, which we can then use to power more complex devices.

In a very simplified system, a hydraulic system is made with piping that has a weight or piston on one end to compress the liquid. As this weight depresses onto the liquid, it forces it out of a much narrower pipe at the other end. The water doesn't squish down and instead pushes itself through the pipe and out the narrow end at high speed. This system works in reverse as well. If we apply a force to the narrow end for a longer distance, it will generate a force capable of moving something much heavier on the other end.

Blaise Pascal, a French mathematician, physicist and inventor, standardized these properties in the mid-1600s. Pascal's Principle states that, in a confined space, any change in pressure applied to a fluid transmits through the fluid in every direction. In other words, if you apply pressure to one end of a container of water, the same pressure will be applied to the other side. This principle is what allows the force to be multiplied and affect a larger, heavier object.

There is a little bit of a trade-off with this system. You can typically apply more force or more speed to one end to see the opposite result on the other. For example, if you press down on the narrow end with high speed and low force, you'll apply high force but low speed to the wide end. The distance your narrow end can travel would also influence how far the wide one will move. Trading distance and force is typical in many systems, and hydraulics are no exception. The multiplication of force is an influential factor in lifting heavy objects. If the piston in the broader side is six times the size of the smaller one, then the force applied to the fluid from the larger piston will be six times as powerful on the smaller end. For example, a 100-pound force down at the wider end creates a 600-pound force up at the narrow end. This force multiplication is what allows hydraulic systems to be relatively small. They are great for powering huge machines without taking up too much space.

Hydraulics can also be very flexible, and there are many different types of hydraulic systems. You can move the fluids through very narrow pipes and snake them around other equipment. They have a variety of sizes and shapes and can even branch off into multiple paths, allowing one piston to power several others. Car brakes are usually an example of this. The brake pedal activates two master cylinders, each of which reaches two brake pads, one for all wheels. You can find hydraulics powering a variety of components through cylinders, pumps, presses, lifts and motors.

Hydraulic systems have a few essential components to control how they work:
Reservoir: Hydraulic systems usually use a reservoir to hold excess fluid and power the mechanism. It is important to cool the fluid, using metal walls to release the heat generated from all the friction it encounters. An unpressurized reservoir can also allow trapped air to leave the liquid, which helps efficiency. Since air compresses, it can divert the movement from the pistons and make the system work less efficiently.

Fluid: Hydraulic fluids can vary, but they are typically petroleum, mineral- or vegetable-based oils. The fluids can have different properties based on their application. Brake fluid, for example, needs to have a high boiling point due to the high-heat mechanism it goes through. Other features include lubrication, radiation resistance and viscosity.

Let's take a look at how hydraulics typically work in heavy equipment:

Engine: This is usually gasoline-powered and allows the hydraulic system to work. In big machines, this needs to be capable of generating a lot of power.

Pump: The hydraulic oil pump sends a flow of oil through the valve and to the hydraulic cylinder. Pump efficiency is often measured in gallons per minute and pounds per square inch (psi).

Cylinder: The cylinder receives the high-pressure fluid from the valves and actuates the movement.

Valve: Valves help to transport the fluid around the system by controlling things like pressure, direction and flow.

Open vs. Closed Hydraulic Systems

Open and closed systems of hydraulics refer to different ways of reducing pressure to the pump. Doing this can help reduce any wear and tear. In an open system, the pump is always working, moving oil through the pipes without building up pressure. Both the inlet to the pump and the return valve are hooked up to a hydraulic reservoir. These are also called "open center" systems, because of the open central path of the control valve when it is neutral. In this case, hydraulic fluid returns to the reservoir. The fluid coming from the pump goes to the device and then returns to the reservoir. There may also be a relief valve in the circuit to route any excess fluid to the reservoir. Filters are usually in place to keep the fluid clean.

Open systems tend to be better for low-pressure applications. They also tend to be cheaper and easier to maintain. One caution is that they can create excess heat in the system if the pressure exceeds valve settings. Another location for added heat is in the reservoir, which needs to be big enough to cool the fluid running through it. Open systems can also use multiple pumps to supply power to different systems, such as steering or control.

A closed system connects the return valve directly to the hydraulic pump inlet. It uses a single central pump to move the fluid in a continuous loop. A valve also blocks oil from the pump, instead sending it to an accumulator where it stays pressurized. Oil remains under pressure but doesn't move unless it is activated. A charge pump supplies cool, filtered oil to the low-pressure side. This step maintains pressure within the loop. A closed system is often used in mobile applications with hydrostatic transmissions and uses one pump to power multiple systems.

These can have smaller reservoirs because they just need to have enough fluid for the charge pump, which is relatively small.

An open system can handle more high-pressure applications. The closed system offers a bit more flexibility than an open system, but that also comes with a slightly higher price tag and more complex repair. Closed systems can work with less fluid in smaller hydraulic lines, and the valves can be used to reverse the direction of the flow.

You can even convert an open system into a closed system by replacing some of the components and adding space for the oil to go after the return trip.

PUMP

Pump, a device that expends energy in order to raise, transport, or compress fluids. The earliest pumps were devices for raising water, such as the Persian and Roman waterwheels and the more sophisticated Archimedes screw (q.v.). The mining operations of the Middle Ages led to development of the suction (piston) pump, many types of which are described by Georgius Agricola in De re metallica (1556). A suction pump works by atmospheric pressure; when the piston is raised, creating a partial vacuum, atmospheric pressure outside forces water into the cylinder, whence it is permitted to escape by an outlet valve. Atmospheric pressure alone can force water to a maximum height of about 34 feet (10 meters), so the force pump was developed to drain deeper mines. In the force pump the downward stroke of the piston forces water out through a side valve to a height that depends simply on the force applied to the piston.

TYPES OF HYDRAULIC PUMPS

There are several different types of hydraulic pumps. These can vary significantly in the ways that they move fluid and how much they displace.

Almost all hydraulic pumps are positive displacement pumps, meaning they deliver a precise amount of fluid. They can be used in high-power applications of over 10,000 psi. Non-positive displacement pumps depend on pressure for the amount of fluid they move, while positive displacement pumps do not. Non-positive pumps are more common in pneumatics and low-pressure applications. They include centrifugal and axial pumps. Positive displacement pumps can have either fixed or variable displacement. Most pumps fall under fixed displacement. In fixed displacement, the pump provides the same amount of fluid in each pump cycle. In variable displacement, the pump can provide different amounts of fluid based on the speed it is run at or the physical properties of the pump. A gear pump is inexpensive and more tolerant of fluid contamination, making them suitable for rough environments. They may be less efficient, however, and wear more quickly.

External gear pumps: These make use of two tight-meshed gears within a housing. One is the driving, or powered, gear, while the other is driven, or free-flowing. The fluid is trapped in the space in between the gears and rotated through the housing. Since it cannot move backward, it is forced through the outlet pump.

Internal gear pump: The internal gear design places an inner gear, possibly with a crescent-shaped spacer, inside of an outer rotor gear. The fluid is moved via eccentricity the deviation of the gear from circularity between the gears. The inner gear, with fewer teeth, turns the outer gear, and the spacer goes in between them to create a seal. The fluid is drawn in, moved through the gears, sealed up and discharged.

NEXT IS VANE PUMPS.

These can be unbalanced or balanced and fixed or variable-displacement. They are quiet and work in pressures under 4,000 psi.

Unbalanced vane pump: This fixed displacement pump has a driven rotor and vanes that slide out in radial slots. The rotor's level of eccentricity determines the level of displacement. As it rotates, the space between the vanes increases, creating a vacuum to draw fluid in. The trapped fluid moves around the system via the rotating vanes and is pushed out as the space between them decreases.

Balanced vane pump: The balanced vane pump, also fixed displacement, moves the rotor through an elliptical cam ring. It uses two inlets and outlets on each revolution.

Variable-displacement vane pump: The displacement in this type of pump can change via the eccentricity between the rotor and casing. The outer casing ring is moveable. Our last category of pumps is piston pumps, which are great for high-powered applications. In-line axial piston pumps: In-line pumps align the center of the cylinder block with the center of the driveshaft. The angle of the swash/cam plate helps to determine the amount of displacement. The inlet and outlet are located in the valve plate, which connects to each cylinder alternately. As the piston moves up past the inlet port, it pulls in fluid from the reservoir. Similarly, it will push the liquid out of the outlet port as it passes it.

Bent-axis axial piston pumps: The bent-axis pumps line the center of the cylinder block at an angle with the center of the drive shaft. This design works similarly to the in-line axial pump.

Radial piston pumps: A radial piston pump uses seven or nine radial barrels, along with a reaction ring, pintle and driveshaft. The pistons are set radially around the drive shaft, and inlet and outlet ports are in the pintle, a type of hinge.

CLASSIFICATION OF PUMPS.

Pumps are classified according to the way in which energy is imparted to the fluid. The basic methods are (1) volumetric displacement, (2) addition of kinetic energy, and (3) use of electromagnetic force.

A fluid can be displaced either mechanically or by the use of another fluid. Kinetic energy may be added to a fluid either by rotating it at high speed or by providing an impulse in the direction of flow. In order to use electromagnetic force, the fluid being pumped must be a good electrical conductor. Pumps used to transport or pressurize gases are called compressors, blowers, or fans. Pumps in which displacement is accomplished mechanically are called positive displacement pumps. Kinetic pumps impart kinetic energy to the fluid by means of a rapidly rotating impeller. Broadly speaking, positive displacement pumps move relatively low volumes of fluid at high pressure, and kinetic pumps impel high volumes at low pressure. A certain amount of pressure is required to get the fluid to flow into the pump before additional pressure or velocity can be added. If the inlet pressure is too small, cavitation (the formation of a vacuous space in the pump, which is normally occupied by liquid) will occur. Vaporization of liquid in the suction line is a common cause of cavitation. Vapour bubbles carried into the pump with the liquid collapse when they enter a region of higher pressure, resulting in excessive noise, vibration, corrosion, and erosion.

The important characteristics of a pump are the required inlet pressure, the capacity against a given total head (energy per pound due to pressure, velocity, or elevation), and the percentage efficiency for pumping a particular fluid. Pumping efficiency is much higher for mobile liquids such as water than for viscous fluids such as molasses. Since the viscosity of a liquid normally decreases as the temperature is increased, it is common industrial practice to heat very viscous liquids in order to pump them more efficiently.

POSITIVE DISPLACEMENT PUMPS.

Positive displacement pumps, which lift a given volume for each cycle of operation, can be divided into two main classes, reciprocating and rotary. Reciprocating pumps include piston, plunger, and diaphragm types; rotary pumps include gear, lobe, and screw, vane, and cam pumps. The plunger pump is the oldest type in common use. Piston and plunger pumps consist of a cylinder in which a piston or plunger moves back and forth. In plunger pumps the plunger moves through a stationary packed seal and is pushed into the fluid, while in piston pumps the packed seal is carried on the piston that pushes the fluid out of the cylinder. As the piston moves outward, the volume available in the cylinder increases, and fluid enters through the one-way inlet valve. As the piston moves inward, the volume available in the cylinder decreases, the pressure of the fluid increases, and the fluid is forced out through the outlet valve. The pumping rate varies from zero at the point at which the piston changes direction to a maximum when the piston is approximately halfway through its stroke. The variation in pumping rate can be reduced by using both sides of the piston to pump fluid.

Pumps of this type are called double acting. Fluctuations in pumping rate can be further reduced by using more than one cylinder.

Overall pumping rates of piston pumps may be varied by changing either the reciprocating speed of the piston rod or the stroke length of the piston. The piston may be driven directly by steam, compressed air, or hydraulic oil or through a mechanical linkage or cam that transforms the rotary motion of a drive wheel to a reciprocating motion of the piston rod.

Piston and plunger pumps are expensive, but they are extremely reliable and durable. Piston pumps are known to have been running without repair or replacement for more than 100 years. The action of a diaphragm pump is similar to that of a piston pump in which the piston is replaced by a pulsating flexible diaphragm. This overcomes the disadvantage of having piston packings in contact with the fluid being pumped. As in the case of piston pumps, fluid enters and leaves the pump through check valves. The diaphragm may be actuated mechanically by a piston directly attached to the dia-phragm or by a fluid such as compressed air or oil. Diaphragm pumps deliver a pulsating output of liquids or gases or a mixture of both. They are useful for pump-ing liquids that contain solid particles and for pumping expensive, toxic, or corrosive chemicals where leaks through packing cannot be tolerated. Diaphragm pumps can be run dry for an extended period of time. Fur-thermore, the pumping rate of most such pumps can be changed during operation. The most common type of gear pump is illustrated in Figure 1. One of the gears is driven and the other runs free. A partial vacuum, created by the unmeshing of the rotating gears, draws fluid into the pump. This fluid is then transferred to the other side of the pump between the rotating gear teeth and the fixed casing. As the rotating gears mesh together, they generate an increase in pressure that forces the fluid into the outlet line. A gear pump can discharge fluid in either direction, depending on the direction of the gear rotation.

An internal gear pump. The driven gear is a rotor with internally cut teeth, which mesh with the teeth of an externally cut idler gear, set off-center from the rotor. The crescent part of the fixed casing divides the fluid flow between the idler gear and the rotor. Gear pumps can pump liquids containing vapors or gases. Since they depend on the liquid pumped to lubricate the internal moving parts, they are not suitable for pumping gases. They deliver a constant output with negligible pulsations for a given rotor speed. Erosion and corrosion lead to an increase in the amount of liquid slipping back through the pump. Since gear pumps are subject to clogging, they are not suitable for pumping liquids containing solid particles. Since they do not need check valves, however, they can be used to pump very viscous liquids.

Lobe pumps resemble external gear pumps, but have rotors with two, three, or four lobes in place of gears; the two rotors are both driven. Lobe pumps have a more pulsating output than external gear pumps and are less subject to wear. Lobe-type compressors are also used to pump gases; each rotor has two lobes. In a screw pump, a helical screw rotor revolves in a fixed casing that is shaped so that cavities formed at the intake move toward the discharge as the screw rotates. As a cavity forms, a partial vacuum is created, which draws fluid into the pump. This fluid is then transferred to the other side of the pump inside the progressing cavity. The shape of the fixed casing is such that at the discharge end of the pump the cavity closes, generating an increase in pressure that forces the fluid into the outlet line.

Screw pumps can pump liquids containing vapors or solid particles. They deliver a steady output with negligible pulsations for a given rotor speed. Since screw pumps do not need inlet and outlet check valves, they can be used to pump very viscous liquids. Although screw pumps are bulky, heavy, and expensive, they are robust, slow to wear, and have an exceptionally long life.

A sliding vane pump. The rotor is mounted off-center. Rectangular vanes are positioned at regular intervals around the curved surface of the rotor. Each vane is free to move in a slot. The centrifugal force from rotation throws the vanes outward to form a seal against the fixed casing. As the rotor revolves, a partial vacuum is created at the suction side of the pump, drawing in fluid. This fluid is then transferred to the other side of the pump in the space between the rotor and the fixed casing. At the discharge side, the available volume is decreased, and the resultant increase in pressure forces the fluid into the outlet line; the pumping rate can be varied by changing the degree of eccentricity of the rotor. Vane pumps do not need inlet and outlet check valves; they can pump liquids containing vapors or gases but are not suitable for pumping liquids containing solid particles. Vane-type compressors are used to pump gases. Vane pumps deliver a constant output with negligible pulsations for a given rotor speed. They are robust, and their vanes, easily replaced, are self-compensating for wear. Pumping capacity is not affected until the vanes are badly worn.

HYDRAULIC TRANSMISSION

Hydraulic transmission, device employing a liquid to transmit and modify linear or rotary motion and linear or turning force (torque). There are two main types of hydraulic power transmission systems: hydrokinetic, such as the hydraulic coupling and the hydraulic torque converter, which use the kinetic energy of the liquid; and hydrostatic, which use the pressure energy of the liquid.

The hydraulic coupling is a device that links two rotatable shafts. It consists of a vaned impeller on the drive shaft facing a similarly vanned runner on the drive shaft, both impeller and runner being enclosed in a casing containing a liquid, usually oil (see figure). If there is no resistance to the turning of the driven shaft, rotation of the drive shaft will cause the driven shaft to rotate at the same speed. A load applied to the driven shaft will slow it down, and a torque, or turning moment that has the same magnitude on both shafts will be developed. In a properly designed hydraulic coupling, under normal loading conditions, the speed of the driven shaft is about 3 percent less than the speed of the drive shaft. By means of a scoop tube, the quantity of liquid in a coupling and the speed of the driven shaft can be varied. Since there is no mechanical connection between the impeller and the runner, a hydraulic coupling does not transmit shocks and vibrations.

The hydraulic torque converter is similar to the hydraulic coupling, with the addition of a stationary vaned member interposed between the runner and the impeller. All three elements are enclosed in a casing containing a liquid, usually oil. The effect of the stationary member is to make the torque, or turning moment, on the driven shaft greater than the torque on the drive shaft.

When the driven shaft is stopped (stalled), the torque on it is a maximum and may be as much as 3.5 times the drive-shaft torque. A hydraulic torque converter acts like an infinitely variable speed transmission, delivering its higher torques when the output speed is low. In automatic transmissions for automobiles, it can be used as a partial or total substitute for a gearbox and clutch.

Hydraulic transmissions of the hydrostatic type are combinations of hydraulic pumps and motors and are used extensively for machine tools, farm machinery, coal-mining machinery, and printing presses. The motor and pump can be widely separated and connected by piping. Such a system, using pressurized water, was built in London in 1882 and is still used to drive machinery to lift bridges and operate hoists.

PASCAL'S PRINCIPLE

Pascal's principle, also called Pascal's law, in fluid (gas or liquid) mechanics, statement that, in a fluid at rest in a closed container, a pressure change in one part is transmitted without loss to every portion of the fluid and to the walls of the container. The principle was first enunciated by the French scientist Blaise Pascal.

Pressure is equal to the force divided by the area on which it acts. According to Pascal's principle, in a hydraulic system a pressure exerted on a piston produces an equal increase in pressure on another piston in the system. If the second piston has an area 10 times that of the first, the force on the second piston is 10 times greater, though the pressure is the same as that on the first piston. This effect is exemplified by the hydraulic press, based on Pascal's principle, which is used in such applications as hydraulic brakes. Pascal also discovered that the pressure at a point in a fluid at rest is the same in all directions; the pressure would be the same on all planes passing through a specific point. This fact is also known as Pascal's principle, or Pascal's law.

12 FASCINATING FACTS ABOUT HYDRAULICS

Hydraulics are one of the technological wonders of the modern world. They are used in a huge variety of different ways, with everything from hydraulic presses, being used to make coins and other items, to the types of Hydraulics used to power the surfaces of planes. Without Hydraulics the world would be a very different place. Here are 12 fascinating facts about Hydraulics, and their influence on our world.

1 -When energy is put into the hydraulic system, it comes out in one of two different ways. It is either work, which is the energy used in the actual operation of the system, or it comes out as loss in the form of heat.

2 -All hydraulic systems must contain four basic components. A reservoir which stores oil, a pump which pumps oil through the system, a valve which controls the pressure and flow of the oil, and a cylinder which converts the fluid movement into work.

3 -Hydraulic presses, like the kind made by hydraulic press manufacturers PressMaster, are used to make coins. Using immense force to compress metals into the right size and

shape.

4 -The hydraulic pumps used in NASA's space shuttle operated at a massive 3600 RPM, and each one could give around 3050 PSI.

5 -Hydraulics will only work if the oil within the system is compressed to an extreme level. Otherwise it simply won't work.

6 -There is no construction of new energy in a hydraulic system. The system just converts existing energy into another form.

7 - Oil is pushed into the system, not pulled by the system itself. This is done entirely by air pressure.

8 -Two hydraulic systems can produce two different flows, from the same amount of energy output. One can be a low flow system, while the other can be a high flow system. Even when the energy is the same.

9 -There are two main types of hydraulic systems. An open center system has a varied pressure but the flow remains constant. In contrast a closed center system has a flow which is varied, but the pressure remains constant.

10 -If the flow encounters any kind of block or opening, then the pressure in the hydraulics will drop, the flow must be unrestricted.

11 -In addition to the two types of hydraulic systems, there are also two basic types of hydraulics. Hydrostatics, which use fluids at low speeds, but at a high pressure to supply power. Most hydraulic systems, like a shop press, are like this. There are also Hydrodynamics which use fluids at a higher speed, but slightly lower pressure, like a

propeller.

12 -There are three main types of hydraulic energy. There is kinetic energy, this is the energy caused by moving liquids. There is heat energy, this is the energy that is the resistance to the flow. Finally there is the potential, or the pressure energy.

CONCLUSION

Other machines that make use of hydraulics include vehicles on construction sites. Diggers, cranes, bulldozers and excavators can all be run by robust hydraulic systems. A digger, for example, powers its massive arm with hydraulic-powered rams. The fluid is pumped into the thin pipes, lengthening the rams and, by extension, the arm. The hydraulic power behind this can be used to lift enormous loads. Aside from construction machines, hydraulics are used for everything from elevators to motors, even in airplane controls.